SCIENTIFIC AMERICAN EDUCATIONAL PUBLISHING

PLAY IT COOL

10 FUN EXPERIMENTS ABOUT TEMPERATURE

BRING SCIENCE HOME

Published in 2024 by The Rosen Publishing Group, Inc.
2544 Clinton Street, Buffalo, NY 14224

Contains material from Scientific American , a division of Springer Nature America, Inc., reprinted by permission as well as original material from The Rosen Publishing Group.

First Edition

Editor: Kristen Rajczak Nelson
Designer: Rachel Rising

Activity on p. 5 by Science Buddies, Ben Finio (February 27, 2020); p. 11 by Science Buddies, Svenja Lohner (February 20, 2020); p. 17 by Science Buddies, Sabine De Brabandere (February 7, 2019); p. 23 by Science Buddies, Svenja Lohner (August 30, 2018); p. 29 by Science Buddies, Svenja Lohner (February 15, 2019); p. 35 by Science Buddies, Svenja Lohner (January 5, 2017); p. 41 by Science Buddies (June 4, 2015); p. 47 by Ben Finio, Science Buddies (April 9, 2015); p. 52 by Exploratorium (December 18, 2014); p. 56 by Science Buddies (August 29, 2013).

All illustrations by Continuum Content Solutions

Photo Credits: pp. 3, 5, 8, 14, 20, 27, 29, 33, 35, 39, 41, 45, 46, 50, 51, 54, 56, 58, 59 cve iv/Shutterstock.com; pp. 4, 5, 11, 17, 23, 29, 35, 41, 47, 52, 56 Anna Frajtova/Shutterstock.com.

Names: Scientific American, Inc.
Title: Play it cool: 10 fun experiments about temperature / edited by the Scientific American Editors.
Description: New York : Scientific American Educational Publishing, 2024. | Series: Bring science home | Includes glossary and index.
Identifiers: ISBN 9781725349186 (pbk.) | ISBN 9781725349193 (library bound) | ISBN 9781725349209 (ebook)
Subjects: LCSH: Temperature--Experiments--Juvenile literature. | Heat--Experiments--Juvenile literature. | Cold--Experiments--Juvenile literature. | Science projects--Juvenile literature.
Classification: LCC QC271.4 P53 2024 | DDC 536.078--dc23

Manufactured in the United States of America

CPSIA Compliance Information: Batch #CSSA24. For further information contact Rosen Publishing, New York, New York at 1-800-237-9932.

CONTENTS

⚛ THESE ACTIVITIES INCLUDE SCIENCE FAIR PROJECT IDEAS.

INTRODUCTION

Temperature affects life every day. It helps us decide what to wear and whether our food is cooked enough. It has many uses in science too! The following experiments all have to do with temperature: how we feel it, what it does to change materials, and how it affects reactions.

Projects marked with ⚛ include a section called Science Fair Project Ideas. These ideas can help you develop your own original science fair project. Science fair judges tend to reward creative thought and imagination, and it helps if you are really interested in your project. You will also need to follow the scientific method. See page 61 for more information about that.

Build a Cooler

KEEP YOUR COOL: USE PHYSICS TO DESIGN THE BEST WAY TO KEEP ICE COLD!

How does a cooler keep things cold? Which material makes the best insulation? Try this project to find out how long you can keep an ice cube from melting once it's out of the freezer!

PROJECT TIME

20 minutes to 1 hour

KEY CONCEPTS

Physics
Heat
Temperature
Insulation

BACKGROUND

Have you used an insulated lunch box or bag to pack lunch for school or a cooler to pack food for a picnic? Why not just throw the food in a paper bag or plain box? What exactly does the cooler do that's so special? You'll find out in this project—but first let's learn a little bit about heat transfer.

Heat naturally flows from hot to cold. You can observe this by watching (and feeling) a hot drink. A cup of hot chocolate will eventually cool down and reach room temperature. In this case, heat flows from the hot drink to the cooler environment. A cup of cold water will also eventually reach room temperature. In this case, heat flows from the warmer environment to the cooler cup.

Heat is the vibrations and collisions between the molecules in a substance. There are three different ways heat can move between objects: conduction, convection, and radiation. Conduction occurs when objects are in direct contact with one another. You feel conduction when you touch, for example, an ice cube or a hot mug. Convection occurs when a fluid (such as air or water) flows over an object. Convection can be natural (for instance, a hot object, such as a pan on the stove, will warm the air around it, causing this warmer air to rise) or forced (such as by a fan in an air conditioner). You feel convection when you stand outside in a chilly winter wind and feel cold. Finally, heat can be transferred by electromagnetic radiation. Most notably, you can feel this when you stand outside in the sun. You can feel the warmth of the sun on your skin—even though the sun is millions of miles away with the vacuum of space between you and it (so there is no matter that allows for conduction or convection)!

That's a lot of different ways for heat to flow from one place to another. That makes keeping food cold (or hot) a challenge. How do you stop heat transfer if it's happening naturally all the time? Or how can you at least slow it down so your food doesn't spoil? The key is insulation! Thermal insulators are materials that help reduce the rate of heat transfer. Some materials are poor conductors of heat so they make good insulators to slow down heat conduction. Other materials can block radiation in visible or infrared light. In this project, you will build your own cooler and explore how insulation can help keep an ice cube from melting (for a while, at least).

MATERIALS

- Two equally sized plastic food storage containers (one will need to have a tight-fitting lid).
- Ice cubes (at least two of the same size)

- Two plastic zip-top bags or matching smaller food storage containers/cups that fit inside the larger ones
- A work surface where both of your containers will have equal exposure to external sources of heat (such as sunlight or a vent). (To speed up the activity put your containers in the sun and/or take them outside on a warm day.)
- An insulating material such as cotton balls, bubble wrap, crumpled paper, and so on (more than one is optional)—enough to mostly fill one of your large food containers

PREPARATION

- Mostly fill the inside of one of your plastic containers with insulating material. This is your "cooler."
- Place a zip-top bag or smaller plastic container inside each of the larger containers (make sure that the lid on the insulated container will be able to close all of the way; remove any insulation needed until the lid can close; leave it open for now).

PROCEDURE

- Place one ice cube in each of your smaller containers or bags.
- Quickly close the lid on the "insulated" cooler container. (Do not put a lid on the other larger container.) *Which ice cube do you think will last the longest?*
- If your workspace isn't very warm, the ice cubes might take a while to melt. Go do something else for 15 minutes, and then come back. *How big is the ice cube in the open container now?*

- Depending on how fast it is melting, continue to check the ice cube periodically. Keep checking until it has melted completely.
- As soon as that ice cube finishes melting, open your cooler and check on that ice cube. *Has it also fully melted? If not, how big is it?*

SCIENCE FAIR IDEA

Try making multiple coolers with different insulating materials—or even combine layers of multiple insulating materials inside a single cooler. *Which material(s) keep the ice cube from melting the longest? Why do you think that is?*

EXTRA

Here's something we haven't considered yet in this project: room! It would probably be nice if you had room for food and drinks in your cooler—instead of filling it mostly with insulation. Try the activity again, but tightly pack the insulation against the walls to leave more room in the container. *Does this change how quickly the ice cube melts? Can you optimize this design with different materials?*

OBSERVATIONS AND RESULTS

When the ice cube in your open container finally melted, you might have discovered that the ice cube in the insulated container was still almost half its original size! Putting the ice cube in a sealed, insulated container dramatically reduces the amount of heat it can gain by conduction (because the insulating material is not a good conductor of heat) and convection (because the lid is sealed, extra air cannot flow around the ice cube). If your insulation is opaque, it also reduces the amount of heat transferred by radiation—although this is a much bigger factor outside in the sun than if you're doing the activity indoors.

Think about what you learned in this activity and how it relates to real coolers. For example, have you ever seen a transparent cooler? Why would that be less desirable? If you want to keep something cold, is it better to leave the cooler lid open or closed? If you need to keep something cold for a long time, would you use a lightweight, thin-walled cooler or a cooler with thicker walls? How does what you learned in this activity impact how you might pack your lunch or a picnic?

CLEANUP

Throw away or let the bags holding the ice cubes dry to use again. Recycle or dispose of your insulation materials. Put away your food containers.

Curl Metals with Heat

BEND METALS IN THIN AIR—AND LEARN HOW A LITTLE PHYSICS CAN HELP US TELL THE TEMPERATURE.

PROJECT TIME

20 minutes

KEY CONCEPTS

Physics
Temperature
Thermal expansion
Metals

Do you enjoy wrapping gifts for people? Perhaps you have even curled a ribbon with scissors (or watched someone else do it). Have you ever wondered why the ribbon curls when you run a scissor blade down one side of it? The answer is that when you apply pressure on the ribbon with the blade of the scissors, the outer layer of the ribbon stretches and expands. This makes the outside layer of the ribbon longer than the inside layer that is pressed against the blade. As a result, the ribbon curls to make up for the different lengths of each of its layers. In this activity, you will also make materials curl. For these materials, however, you won't need scissors: you will use heat instead!

BACKGROUND

You probably know or have seen that materials are able to change their shape or volume when they are heated or cooled. This is true for solids, liquids, and gases, which are all made up of atoms and molecules. Once exposed to heat, these molecules begin to vibrate and move around faster. This makes the molecules spread out and take up more space. As a result, a material expands slightly when it gets hot. On the other hand, when a material is cold, molecules move less, which makes them take up less space. Thus, materials shrink when they get cold. Although all materials expand when heated, they do not all expand to the same degree. How much a material expands when heated is described by its thermal expansion coefficient. For example, aluminum expands 21 to 24 micrometers per meter if you increase its temperature by 1 degree Celsius.

What happens when an object is made up of more than one material? They will both expand differently when heated up! In fact, there are special materials called bimetals that make use of their different thermal expansion properties. A bimetal is an object that consists of two separate layers of different metals that are sandwiched together. When a bimetal is heated, one of the metals will expand more than the other. This results in the bimetal curving (or curling) in one direction—just like the ribbon for the gift wrap. Because of this effect, bimetals are often used to indicate temperature changes, such as in pointer dial thermometers (such as those used in ovens or refrigerators). Inside these thermometers a bimetal coil is attached to a pointer. When the temperature changes, the pointer moves depending on the amount of deformation of the bimetal coil.

In this activity, you will see for yourself how different thermal expansion properties can make an object curl. Ready to bring on the heat?

MATERIALS

- Aluminum foil
- Printer paper
- Scissors
- Glue
- Candle with a sturdy base
- Lighter or matches
- Adult helper
- Large pitcher of water or fire extinguisher
- Other material sheets (plastic, copper, etc., are optional)

PREPARATION

- Cut four 0.5 by 9-inch (1.2 by 23 cm) strips of aluminum foil.
- Cut four 0.5 by 9-inch (1.2 by 23 cm) strips of printer paper.
- Glue two of the aluminum strips together, so there are two layers of aluminum on top of each other.
- Glue two of the paper strips together, so there are two layers of paper on top of each other.
- Glue one strip of paper on top of one strip of aluminum.
- Repeat the previous step with the remaining aluminum and paper strip.
- Ensure the glue is set before beginning the activity.

PROCEDURE

- With the help of an adult, light the candle. Ensure that you have a large pitcher of water or fire extinguisher nearby—just in case.
- Take the strip that is made up of two layers of aluminum. With the help of an adult, carefully hold one end of the strip about 1.5 to 2 inches (4 to 5 cm) above the candle flame. Keep it there for about three seconds. *What happens to the aluminum strip?*
- Remove the aluminum strip from above the candle. Carefully touch the part of the strip that was above the flame. *How does it feel compared to the rest of the strip?*
- Take the strip that is made up of two layers of paper. Hold one end of the strip about 1.5 to 2 inches (4 to 5 cm) above the candle flame. Keep it there for three seconds, being careful to not accidentally burn the paper strip. *What do you notice this time?*

- Remove the paper strip from above the candle, and gently touch the part of the strip that was above the flame. *How does it feel?*

- Next, take a strip that is made up of aluminum and paper. Hold one end of the strip above the candle flame as you did with the others. Have the aluminum layer face toward the candle and the paper layer toward the ceiling. Keep the strip 1.5 to 2 inches (4 to 5 cm) above the candle for three seconds. *What do you observe? Are your results different than before?*

- After removing the strip from above the candle, carefully touch the strip on both sides where it was above the flame. *Do both sides feel the same?*

- Take the last strip made up of aluminum and paper. This time, hold the strip above the flame with the paper layer facing the candle flame and the aluminum layer facing up. *What happens this time? Can you explain your results?*

- Remove the strip from above the candle and carefully touch it on both sides where it was above the flame. *What do you notice?*

EXTRA

What happens when you glue the aluminum strip to other materials such as a plastic sheet or copper sheet? Do your results change?

OBSERVATIONS AND RESULTS

Did you see one of your strips curl above the candle flame? You should have! The all-aluminum strip and the all-paper strip should have not changed when you held them above the flame. This is because the two layers of the strip are made of the same material. If the strip gets hot, which you should have noticed when touching it, both layers expand the same amount. If the strip consists of two different materials, however, such as the one you made from paper and aluminum, something different will happen.

When holding the paper/aluminum strip above the flame, you probably observed that it immediately started to bend or curl in one direction. It curls upward when the aluminum layer faces the candle and downward when the aluminum layer faces the ceiling. This is because the aluminum and the paper layer heat up when above the flame. The aluminum expands more than the paper because of its higher thermal expansion coefficient. Because the paper and aluminum are glued together, the aluminum curls away from the paper strip to make up for the different lengths of each of its layers—like the curling gift wrap ribbon! You will see similar results with other strips that consist of two different materials that have different thermal expansion coefficients.

CLEANUP

Make sure to blow out the candle. You can recycle any unused paper or aluminum.

Stretchy Balloon Science

STRETCH IT! FIND OUT HOW FAR YOU CAN BEND A BALLOON'S SURFACE BEFORE IT REACHES ITS BREAKING POINT.

Balloons are fun and come in a variety of sizes, colors and forms that make them well suited for decorations, water play, modeling, and other creative activities. But balloons also have uses outside of recreation: monitoring weather, widening obstructed arteries and providing transportation are just a few examples. The first balloons were made of animal bladders or intestines. This made way for the latex, rubber, and nylon fabric balloons we see today. Have you ever wondered what makes these materials excellent for balloons? Try this activity to find out!

PROJECT TIME

2 hours of freezing time, 20 minutes of activity

KEY CONCEPTS

Physics
Materials
Elastic
Temperature

BACKGROUND

The stretchiness of a balloon is based on its material. Most balloons are made of rubber, latex, or nylon fabric, which consist of long particles called polymers that are like strands of cooked spaghetti, but a lot smaller! When you stretch these materials, the polymers in them straighten out and slide alongside and over one another. When you let go, small forces between the polymers pull them back together, like a contracting spring. Because of this, these materials are called "elastic," which means they spontaneously return to their normal shape after they are stretched or compressed.

When you inflate a balloon, the trillions of polymers in its material are straightening and sliding over one another: The first puff of air straightens them, and after that each blow works to push the polymers over one another, working against the forces pulling them back together. If you forget to trap the air inside the balloon by pinching it closed, the polymers pull back together. This pushes the air out of the balloon, and you will have to start over!

Balloons are elastic—but only to a certain extent. When you stretch them, they will generally regain the original form after you let the air out. If you stretch them too much, however, they will never quite regain original form. If you stretch them even more, you will eventually reach the breaking point and the material will rupture.

The temperature of matter is an indication of how much its particles move and wiggle around. Because higher temperatures mean more movement, the polymers in a room-temperature balloon wiggle and move more than do those in a cold one. A lot of materials become brittle and inflexible when cooled. A slice of bread is a good example: It is flexible and sometimes elastic at room temperature but loses these characteristics when frozen. Would a cold balloon resist being stretched too?

MATERIALS

- Several new balloons, all the same size and material
- Permanent marker
- Freezer
- Skewer
- Lip balm or petroleum jelly

PREPARATION

- Mark two uninflated balloons with permanent marker.
- Cool the two marked balloons in the freezer for at least two hours. Leave the other balloons at room temperature.

PROCEDURE

- Take a room-temperature balloon, and blow a breath of air into it. *How hard is it to blow air into the empty balloon?*
- Blow a few more breaths into the balloon. *Is it getting easier or harder to push air into the balloon? Why would this be?*
- Let the air escape from the balloon and take some deep breaths before continuing.
- Take a balloon out of the freezer and repeat the previous three steps with this balloon. *How hard is it to blow up an ice-cold balloon compared with one at room temperature?*
- To make a fair comparison, repeat previous steps, now testing the other ice-cold balloon first and the room-temperature balloon last. *Do your results change? Why would one be more difficult than the other?*
- Take one of the room-temperature balloons you just used and inflate it until you are afraid it will burst if you blow any more air into it. Then let the air escape.
- Lay a new balloon next to the two other used room-temperature balloons. *In what ways do these three empty balloons look similar and different?*
- Rub some lip balm on the skewer, starting from the tip.

- Blow up one of the balloons (any of them is fine) until about two-thirds full, and knot it so the air is trapped. *What do you think will happen when we poke the balloon with the skewer?*

- Look at your blown-up balloon. *Where would you poke the skewer into the balloon to pop the balloon quickly? Why do you choose that spot? Can you find a spot where you might be able to poke the balloon with the skewer without popping it?*

- Poke the balloon to pop it. *Did it pop? What do you think creates the loud sound of a popping balloon?*

- Inflate another balloon until it is about two-thirds full. Knot it and examine it. *If you wanted to poke the skewer into this balloon without popping it, where would you poke it?*

- Try it out. *Were you successful?*

- If it did not work, try with another balloon. Notice how the rubber near the knot or the top looks darker and thicker. These are indications the rubber can still stretch there. Try it a few more times until you succeed in poking your skewer through the balloon without popping it. *Why do you think the skewer can enter the balloon in these spots without popping it?*

EXTRA

This activity compares ice-cold and room-temperature balloons. Add a warmed-up balloon to the comparison. You can warm balloons in a warm water bath.

EXTRA

Inflate two new balloons to the same size. Knot them carefully so no air escapes. Cool one balloon in the refrigerator, the freezer, or outside on a freezing-cold day, keeping the other balloon at room temperature. After about an hour, compare the two balloons. *How are they different? Why does this happen?*

OBSERVATIONS AND RESULTS

Did the cold balloon require more effort to blow up compared with the one at room temperature? Did your balloon pop when you poked it in a spot where the balloon was stretched taut?

The particles in cold material move and wiggle less and can hold onto one another more easily. As a result, a cold balloon resists being stretched more than a warmer one does. You perceived this as having to work harder to blow up the cold balloon. You might have noticed this resistance disappears as the balloon warms up.

Did you observe how the balloons that were blown up with only a few breaths regained their original shapes and sizes when the air was let out, but the one that was blown up completely did not quite regain its original size? This is typical for elastic materials. If you had blown even more air into the balloon, it would have eventually broken.

When you tried to pop a taut part of the balloon by poking it, it probably cracked. Think of a stretched rubber band. When you cut it in the middle, the two ends snap back. In a similar way, the small hole in the taut surface made part of the surface snap back, pulling the material around back. This pull tore the surface of the balloon open. The air in the balloon rushed out, creating a shock wave, which you heard as a loud bang.

When you use a sharp object to poke in a place where the material can still stretch, such as close to the knot or near the top, you can pierce a balloon without destroying it. When the material around the puncture is not stretched, it does not pull on the created hole and thus does not tear it open. (Also, the greasy lip balm you applied to the skewer helps to seal the hole.)

CLEANUP

Throw out the broken balloon pieces and any balloons you blew air into. Return other materials to where you found them.

Turn the Sun's Energy into Power

CAN YOU USE A LAMP TO TWIRL A TURBINE? USE A LITTLE BIT OF PHYSICS TO SPIN UP AN ALTERNATIVE FORM OF ENERGY HARVESTING.

Have you ever had a sunburn? If so, you know how powerful the sun's heat can be! Did you know this heat can also be turned into electrical energy? You might have heard about solar cells, but what about a solar updraft tower? This very simple structure uses the sun to heat up air, which then powers a turbine within a large tower. In this activity, you will build one of these towers yourself. Do you think you can make a propeller spin with just hot air?

PROJECT TIME

45 minutes

KEY CONCEPTS

Physics
Energy
Density
Temperature
Greenhouse effect

BACKGROUND

The sun is a very powerful source of energy. In fact, in less than 15 seconds it can provide as much energy to Earth as humans use in one day! Most of this energy reaches Earth in the forms of light and heat. It's no wonder people have long considered how this energy can be harvested and put to good use! During the past two centuries, scientists researched and developed technologies to convert the sun's energy directly into electrical energy. Today, many different methods of solar power generation exist, such as solar cells, solar-thermal collectors, and solar-concentration systems.

One of these systems, called the solar updraft tower, is solar-thermal collector technology. A solar updraft tower can generate electricity from low-temperature solar heat. The science concepts behind this idea are based on the greenhouse effect and the chimney, or stack, effect. These power plants consist of a large collector structure and a very tall and slender tower. The collector structure resembles a gigantic greenhouse and consists of a massive, transparent canopy that is suspended 6 to 65 feet (2 to 20 m) off the ground. The tower is located in the middle of the canopy structure and has large air inlets at its base. Inside the tower are big wind turbines to produce electricity.

When the sun shines onto the collector structure, its heat gets trapped underneath as in a greenhouse, which warms the air and ground underneath the canopy structure. As hot air is less dense than cold air, it starts to rise, causing an airflow, also called thermal convection. The air has no other way to escape than through the tower. The rising air in the tower creates low pressure at the bottom of the tower, and more hot air from the collector is sucked inside the tower through its air inlets at its base. As a result, solar radiation causes a continuous updraft in the tower. The wind turbines installed inside the tower convert the energy contained in the updraft into mechanical energy by moving its blades. In the next step, this movement is converted into electrical energy by generators.

Although the technology seems pretty simple and can generate lots of energy, no full-scale units are in operation—yet. Several prototype models, however, have been built and studied. In this activity you will build your own very simple solar updraft tower prototype—but without the collector structure. Can you use the energy created by the updraft to power a propeller? Get started to find out!

MATERIALS

- Black construction paper (about 20-by-20 inches [51-by-51 cm])

- Pencil
- Scissors
- Tape
- Piece of modeling clay
- Wooden skewer
- Needle
- Thermometer
- Lamp with incandescent lightbulb or a heat lamp (An LED bulb is too cool to produce enough heat.)
- Sunshine and an area outside protected from wind (optional)
- Stopwatch (optional)
- White construction paper (optional)

PREPARATION

- Roll the black construction paper into a cone with a small opening at the top (2-inch [5 cm] diameter) and a larger opening at the bottom (4-inch [10 cm] diameter).

- Use tape to hold the cone in place. Then cut the top and bottom ends off to make them straight. The cone should be about 10 to 15 inches (25 to 38 cm) high and able to stand by itself.

- At the bottom of the cone, cut three equally distanced 2-inch by 1/2-inch (5 cm by 1 cm) arches out. These will be your air inlets. Make sure the cone can still stand on the remaining rim.

- Cut a 3-inch (7.6 cm) diameter propeller out of the construction paper. You can find propeller templates online. Bend the blades of the propellers at about a 45-degree angle downward.

- With tape, carefully attach the needle to the top of the wooden skewer with the pointy end facing upward.

- Make a ball out of modeling clay, and press it on your workspace.

PROCEDURE

- Look for a wind-protected space indoors to set up your solar updraft tower. It is important there is no external airflow happening during your test. *Why would your workspace need to be wind-protected?*

- Place the black paper cone over the clay so that the clay is located in the center of the cone.

- Stick the wooden skewer into the clay through the top opening of the tower so the skewer sticks out in the middle of the cone. The needle on top of the skewer should stick about 1 1/2 to 2 inches (4 to 5 cm) out of the cone.

- Take the propeller and push it onto the point of the needle. After you mount the propeller, it should be balanced and able to turn freely. You might need some practice to make sure the propeller is resting on the needle correctly. *What happens to the propeller resting on top of the needle? Does it move?*

- Measure the air temperature inside the tower and outside. *What inside air temperature do you measure compared with the outside temperature?*

- Place the lamp next to the tower and direct it toward the base of the tower. Make sure you don't leave the lamp unattended during the test. *What do you think the lamp will do?*

- Switch on the lamp and observe the updraft tower for a while. *What happens?*

- After about five minutes, measure the air temperature inside and outside of the cone again. *What temperatures do you measure this time?*

- Switch the lamp off and continue to observe your updraft tower for five to 10 minutes. *What happens when you switch off the lamp?*

- After another five minutes, measure the temperature inside and outside the cone one more time. *Did the temperatures change? If yes, how?*

EXTRA

How fast can you make your propeller spin? Mark one of your propeller blades with a pencil and then count the number of turns in 15 seconds using a stopwatch. Multiply the counted turns by four and you get the rotation speed of your propeller in rpm (rotations per minute).

EXTRA

Repeat the test, but this time use white construction paper to build your solar updraft tower. *Does the color of the paper make a difference? Why or why not?*

EXTRA

Go outside and let the sun power your solar updraft tower instead of the lamp. Make sure you find a wind-protected spot. *Can the sun power your propeller too?*

OBSERVATIONS AND RESULTS

Did you get your propeller to spin? In the beginning the propeller shouldn't have moved. If it did, an external airflow from a door or wind was likely moving the propeller. The air temperatures inside and outside the tower should have been pretty similar. This should have changed, however, when you switched on the lamp. The incandescent lamp emits light and heat, just like the sun. You will notice this when holding your hand close to the lightbulb. When the light shines onto the black construction paper, it absorbs most of the light and starts to warm. The heat is trapped inside the cone, which increases the air temperature inside. Because warm air is less dense than cold air, it starts to rise up the paper cone. This updraft makes the propeller on top of the cone spin, which you should have observed after a while.

The rising air reduces the air pressure inside the cone, which is why fresh air is sucked into the paper cone through the air inlets that you cut out at the bottom. This new air is warmed, and a continuous updraft is created that keeps the propeller spinning. The cycle is only interrupted when you switch off the lamp. Because there is no heat source, the air inside the cone slowly starts to cool, and after a while returns to the same temperature as the outside air. At that point, the propeller should have stopped spinning because there was no longer an updraft. You might have noticed it took longer for the propeller to spin and the propeller didn't spin as fast with the white construction paper. This is because white paper does not absorb the heat from the lamp as well as the black paper can.

CLEANUP

Carefully remove the needle from the skewer. You can reuse all the parts of your construction.

Find Out Why the Ocean Is in Motion!

MAKE SOME UNDERWATER WAVES—WITH CONVECTION CURRENTS!

PROJECT TIME

40 minutes

KEY CONCEPTS

Physics
Convection
Temperature
Density

You probably know the ocean never really stays still. But did you know there is something called the global "ocean conveyor belt" that moves massive amounts of water from one ocean to another? These water currents are essential for mixing and transporting nutrients and oxygen and play a critical role in our climate. This is because they move warm and cold water over very long distances, which affects the temperature of the landmasses that border the ocean. The Gulf Stream, for example, is a large current flowing from the Gulf of Mexico that moves warm water into the northern Atlantic Ocean and makes Europe's climate much milder. Do you know what causes the "ocean conveyer belt"? It is pretty simple: differences in water density. Try this activity to learn more!

BACKGROUND

There is a lot of water movement in the ocean. The most obvious examples are the waves and ripples on the water's surface that are generated by wind or the ocean currents that are due to tides. It turns out, however, that water can also be moved without wind or tides, which is what happens in the deep ocean. There, currents are set in motion by variations in water density caused by differences in temperature and salinity, a process called convection.

The density of a substance is defined by its mass per volume; 1 gallon (3.8 l) of water, for example, does not weigh the same as 1 gallon (3.8 l) of oil. The lighter liquid floats on the heavier liquid, which sinks to the bottom. Water's density varies with temperature; when heated it will expand and therefore increase in volume. This means its mass per volume will decrease and it becomes less dense. Warm water, thus, is generally less dense than cold water. Another factor that affects water density is salinity, which is the amount of salt in ocean water. More salt makes water heavier and therefore denser.

These two factors, temperature and salinity, are the main driving forces behind Earth's ocean conveyor belt, which is a huge water circulation system in the deep ocean that moves water around the globe. Currents begin near the North Pole in the North Atlantic where the surface of the ocean gets cooled by the Arctic temperatures. As sea ice forms, salt stays behind and the water there becomes saltier. This water is now denser and will sink to the bottom of the ocean floor. This creates a current as surface water has to move in to replace the sinking water. The deep, cold water moves all the way to Antarctica and then to the Indian and Pacific Oceans. Once the water reaches warmer regions, it heats up, becomes less dense, and rises toward the surface. It eventually finds its way back to the North Atlantic where the whole cycle begins again. In this activity, you will demonstrate how temperature affects water movement and water density—and create your own little ocean currents in a glass!

MATERIALS

- Two transparent cups, exactly the same size
- An old CD
- Sandwich bag
- Scissors
- Food coloring (any color)
- Tap water
- Microwave oven

- Ice cubes
- Thermometer
- Paper towels for water spills
- Adult helper
- Tray
- Table salt (optional)

PREPARATION

- Use the scissors to cut a long strip off the sandwich bag, wide enough to fully cover the hole in the middle of the CD.
- Place the tray on a work area that can tolerate water spills.
- Fill both cups with tap water.
- Add ice cubes to one of the cups to cool down the water. Once it is cold, remove the ice cubes and fill the cup to the brim with more tap water.
- Put the other cup into the microwave for about one minute. (If it is too hot, let it cool down until you can pick up the cup.)
- Make sure it is filled to the brim with water.
- Add a couple of drops of food coloring to the hot water.
- Set the cup with hot water into the tray.
- With the thermometer, measure the temperature of both the hot and cold water.

PROCEDURE

- Place the cut plastic strip from the sandwich bag onto the CD so the hole is completely blocked, and part of the plastic strip extends beyond the CD.
- Put the CD on top of the cold cup with the plastic strip facing toward the cold water.

- Then pick up the cup with cold water and the CD on top and slowly turn it upside-down, pressing on the CD to avoid any spillage. You might need an adult to help you with this step.
- Place the upside-down cup with cold water on top of the cup with hot water so the CD is separating both cups. *Once you have placed both cups together, what happens? Do you see any water mixing yet?*
- Slowly pull on the part of the plastic strip that extends beyond the CD and remove the whole strip to expose the CD's hole. *What happens when the hole is open? Do you see any water movement? If yes, in which direction does the water move? Can you describe your observations?*
- Observe both cups for about 10 minutes. *Do you see any changes happening in the top or bottom cup? If yes, how does the water change?*
- After 10 minutes, carefully separate both cups again by lifting the CD and the upper cup and inverting both again. You might want to ask an adult to help you with that.
- Finally, measure the temperature again in both cups. *Has the temperature changed? If yes, how?*
- Drain the water in both cups into the sink and set up the same test again, but this time place the cup with hot water on top of the cup with cold water. (You should have an adult to help you with this.) *What happens if you remove the plastic strip this time? Do you see the same results? Does any water movement happen? Why or why not?*
- After 10 minutes, separate both cups again and measure the temperature in each cup. *Has the temperature changed? If yes, how much?*

EXTRA

In this activity, you have shown differences in temperature drive water movement. *What happens if you have no temperature gradient?* Repeat the same test, but this time do not heat or cool the water in the cups—keep them at the same temperature. *Do you see any water movement happening when you remove the plastic strip? Why or why not?*

SCIENCE FAIR IDEA

Can you find a way to demonstrate how salinity affects water density and water movement? Add salt to one of the cups to create a salinity gradient and repeat the test. *Does the salty water float or sink? Can you create water movement with a salt gradient?*

EXTRA

Instead of demonstrating that warm water floats atop cold water, show how cold water sinks to the bottom and add the food coloring to the cold water instead of the warm water. Then place the cold water on top of the warm water and vice versa. *Do you see water movement in both directions, or only in one?*

SCIENCE FAIR IDEA

How does the temperature gradient affect the water mixing? Add a third test in which the temperature difference between the warm and cold water is different from your previous test. *What changes if you have a higher versus a lower temperature difference between the hot and cold water?*

OBSERVATIONS AND RESULTS

Were you able to make your water move to float and sink? When you add the food coloring to the hot water and place the cold water on top of the hot water, initially nothing should have happened as the plastic strip should have still covered the hole of the CD, preventing the water from mixing between the cups. As soon as you remove the plastic strips, however, the dyed warm water will shoot through the CD hole and mix with the cold water above.

This water movement is driven by convection: The movement of water due to differences in temperature. Because the warm water is lighter than the cold water and wants to expand, it moves into the upper cup through the hole. After about 10 minutes, the mixing should almost be completed, and the color of the water should be similar in both cups. At the same time, the temperature in both cups should have changed significantly. In the beginning you should have measured a large temperature difference. But at the end, both temperatures should have been similar. As the water was mixing, heat exchange occurred, which means the warm water heated the cold water. If you tried the extra activity with a higher or lower temperature difference between both cups, you should have noticed the mixing of water happens faster when the temperature difference is larger. When you have no temperature gradient at all, no significant mixing will happen within 10 minutes.

If you reversed the test and have the warm on top of the cold water, you should have observed the water did not mix between the cups. This is because warm water is lighter, or less dense, than cold water, which is why it wants to remain on top. The same is true for salty versus non-salty water. Salty water will always sink to the bottom, as it is denser than non-salty water—exactly how it does in the real ocean!

CLEANUP

Pour all the water into the sink and clean any spills.

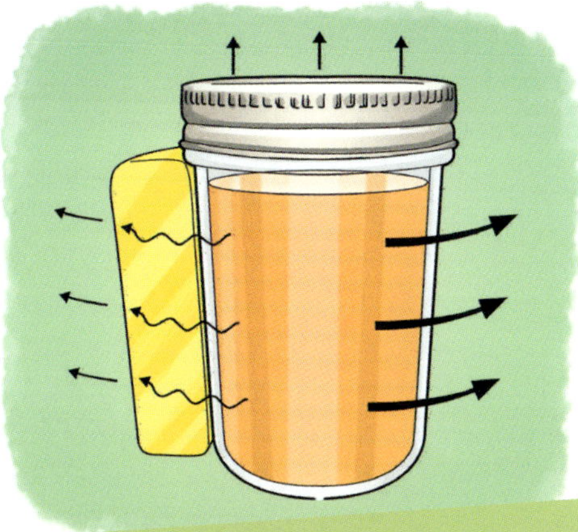

Stay Warm with Thermal Insulation

WANT TO KEEP WARM THIS WINTER? TRY THIS "COOL" ACTIVITY AND FIND OUT WHAT TYPES OF INSULATION WORK BEST—AND WHY.

What do you do when it gets very cold in winter? You probably turn your heater on, put on an extra layer of clothes, or cuddle under a warm blanket. But have you ever thought about why a jacket helps you stay warm? Why are our clothes made from fabrics and not foils? Find out the answers in this activity; your results might even help you find the best way to stay warm in the cold!

PROJECT TIME

30 minutes

KEY CONCEPTS

Physics
Heat transfer
Insulation
Material science

BACKGROUND

Heat is a form of energy. You need energy to heat something up: for example, a cup of tea. To make your tea you probably use energy from electricity or gas. However, once your tea is hot, it won't stay hot forever. Just leave the cup of tea out on the table for a while, and you already know that it will become cooler the longer you wait. This is due to a phenomenon called heat transfer, which is the flow of energy in the form of heat. If two objects have different temperatures, heat automatically flows from one object to the other once they are in contact. The heat energy is transferred from the hotter to the colder object. In the case of the tea, the heat of the liquid is transferred to its surrounding air, which is usually colder than the tea. Once both objects reach the same temperature, the heat transfer will stop. Heat transfer via the movement of fluids (liquids or gases) is called convection.

Another type of heat transfer is conduction, in which energy moves through a substance (usually a solid) from one particle to another (unlike in convection where it's the heated matter itself that moves). A pot handle getting hot would be an example of conduction.

Heat can also be transferred through radiation. You might have experienced that from sitting around a bonfire. Although you are not touching the fire, you can feel it radiating heat in your face even if it is cold outside. If you like to drink your tea hot, you might ask how heat transfer can be reduced, and how the tea keeps from cooling down. The answer is thermal insulation. Insulation means creating a barrier between the hot and the cold object that reduces heat transfer by either reflecting thermal radiation or decreasing thermal conduction and convection from one object to the other. Depending on the material of the barrier, the insulation will be more or less effective. Barriers that conduct heat very poorly are good thermal insulators, whereas materials that conduct heat very well have a low insulating capability. In this activity, you will test which materials make good or bad thermal insulators with the help of a glass of hot water. Which material do you think will be most effective?

MATERIALS

- Five glass jars with lids
- Scissors (and adult to help with cutting)
- Tape
- Aluminum foil
- Bubble wrap

- Wool scarf or other wool clothes
- Paper
- Hot tap water
- Thermometer
- Fridge
- Timer
- Paper for writing
- Pen or pencil

PREPARATION

- Cut a piece of the aluminum foil, the bubble wrap, and the paper (have an adult help if necessary). Each piece should be large enough to fit three times around the sides of the glass jar.

- Take the piece of aluminum foil and wrap it around the sides of one of the jars. You should have three layers of foil around the glass jar. Use the tape to attach the foil to the jar.

- Next, wrap another jar with the bubble wrap so that the glass is also covered in three layers. Make sure to tape the bubble wrap to the jar.

- Use the cut paper to wrap a third jar in three layers of paper. Once again, attach the paper to the glass jar.

- Take another glass jar and wrap the scarf or other wool fabric around the jar. Only make three layers of wrapping and make sure that the scarf stays attached to the jar.

- Leave the last jar without any wrapping. This will be your control.

PROCEDURE

- Fill each jar with the same amount of hot water from your faucet.

- Use the thermometer to measure the temperature in each jar. Put your finger inside the water of each jar (use caution if your tap water is very hot). *How does the temperature of the water feel?*

- Write down the temperature for each jar and close the lids. *Are all the temperatures the same or are there differences? How big are the differences?*

- Open your fridge and put all the five jars inside. Make sure they are still securely wrapped. Feel the temperature of the fridge. *What does its temperature feel like?*

- Put the thermometer in the fridge. *What temperature does the thermometer read when you put it into the fridge?*

- Once all the jars are in the fridge, close the fridge door and set your timer to 10 minutes. *What do you think will happen with the jars and the hot water during that time?*

- After 10 minutes, open the fridge and take all the jars out. *Do the jars feel different?*

- Open each jar, one at a time, and measure the water temperature with your thermometer. Also, feel the temperature with your finger. *Did the temperature change? How did it change according to the thermometer?*

- Repeat measuring the temperature for each jar and write down the temperature for each wrapping material. *Did the temperature in each jar change the same way? Which wrapping material resulted in the lowest temperature change, and which resulted in the biggest?*

- For a better comparison, calculate the temperature difference from the beginning and end of the test for each jar (temperature beginning versus temperature after 10 minutes in fridge). *From your results, can you tell which material is the best or weakest thermal insulator?*

EXTRA

Will temperatures continue to change in a similar way for each material? You can close each jar again and put them back into the fridge for another 10 minutes. *Are the results different this time or the same?*

EXTRA

Does the water temperature change the same in the fridge as in the freezer or at room temperature? Repeat the test, but this time instead of putting the glass jars into the fridge, put them into the freezer or keep them at room temperature. *How much does the temperature of the water change within 10 minutes? Do the different wrapping materials behave differently?*

SCIENCE FAIR IDEA

Try to find other materials that you think are good or bad thermal insulators and test them. *Which material works the best? Can you think of a reason why?*

EXTRA

If you take the jars out of the fridge after 10 minutes, you probably still measure a temperature difference between the water inside the jar and the temperature inside the fridge. You can keep the glass jars longer in the fridge and measure their temperature every 15 to 30 minutes. *How long does it take until the temperature of the water doesn't change anymore? What is the end temperature of the water inside the glass?*

SCIENCE FAIR IDEA

Besides choosing the right insulator material, what are other ways to improve thermal insulation? Repeat this test with only one wrapping material. This time change the thickness of your insulating layer. *Do you find a correlation between thickness of insulation layer and temperature change in the fridge?*

OBSERVATIONS AND RESULTS

Did your hot water cool down significantly during the 10 minutes inside the fridge? Although the fridge temperature is very low, your hot water has a high temperature. As heat energy flows from the hot object to the cold object, the heat energy from your hot water will be transferred to the surrounding cold air inside the fridge once you put the glass jars inside. The most significant mechanism of heat transfer in this case is convection, which means that the air just next to the hot jar is warmed up by the hot water. Then, the warm air is replaced with cold air, which is also warmed up. At the same time, the cold air cools down the water inside the jar. The heat of the hot water is transported away by the flow of cold air around the cup. If you left the jars in the fridge long enough, you might have observed that the temperature changes until the hot water reaches the temperature inside the fridge. Without a temperature difference between the water and the fridge, the heat transfer will stop.

Heat from the water is also lost through conduction: the transfer of heat through the material, which is dependent on the thermal conductivity of the material itself. The glass jar can conduct heat relatively well. You notice that when you touch the glass jar with the hot water the glass feels hot as well. What effect did the different wrapping materials have? You should have noticed that with wrapping materials, the temperature of the water after 10 minutes inside the fridge was higher compared to the unwrapped control. Why? Wrapping the glass jar reduces the heat transfer from the hot water to the cold air inside the fridge. Using wrapping materials that have a very low thermal conductivity reduces the heat loss through conduction. At the same time, the insulator can also disrupt or reduce the flow of cold air around the glass jar, which results in less heat loss via convection.

One way of reducing convection is creating air pockets around the jar, for example, by using insulators such as bubble wrap, fabric, or wool, which have a lot of air pockets. Air in general is a good thermal insulator, but it can transmit heat through convection. However, if the air pockets inside the insulating material are separated from each other, heat flow from one air pocket to another cannot happen easily. This is the reason why you should have measured the highest temperature in the bubble-wrapped jar and fabric-wrapped jar. This also explains why most of our clothes are made of fabrics and why you stay warmer when you put on an extra jacket. Paper and foil make it easier for the heat to escape because they don't have many air pockets.

CLEANUP

Use your leftover water to make tea or water houseplants. Recycle or throw out the insulating materials, or find another use for them. Put back all other materials where you found them.

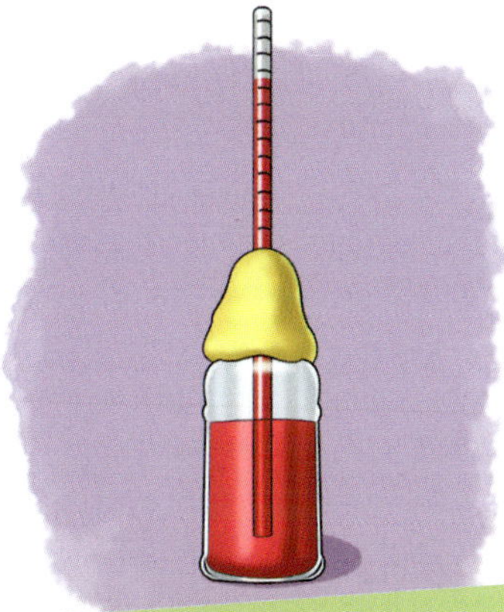

Measure Up with a Homemade Thermometer

IS YOUR HOME HEATING UP THIS SEASON? COOLING DOWN? YOU CAN TRACK TEMPERATURE DIFFERENCES INSIDE AND OUT WITH A THERMOMETER YOU BUILD YOURSELF!

Are you sometimes eager to understand how things work? Or excited about making useful objects and instruments yourself? Did you ever imagine you could build your own liquid thermometer? You'll be able to use it to track how temperatures vary with location—indoors or outdoors. What will turn out to be the hottest spot in your home? What about the coolest? Your very own homemade thermometer will be able to tell you!

PROJECT TIME

30 minutes

KEY CONCEPTS

Temperature
Measurement
Thermal expansion

BACKGROUND

Liquid-filled thermometers have been used for centuries to measure temperature. They consist of a liquid-filled reservoir at the end of a narrow tube. If the temperature of the liquid in the reservoir increases, it expands and rises into the narrow tube. When the liquid cools, it contracts, allowing the liquid in the tube to fall to a lower level. Therefore, high levels of liquid in the tube indicate a higher temperature and low levels indicate a lower temperature. When calibrating the thermometer, the specific temperatures corresponding to the levels of the liquid in the narrow tube are marked on the outside of the tube. Now you understand how a liquid thermometer works, but why does it work?

Scientists refer to the tendency of matter to change volume as a result of a temperature change as "thermal expansion and contraction." Gases expand and contract drastically. Fluids expand and contract, too, but their volume change is more moderate. Even solids change volume when they are heated or chilled. For example, bridges are a little bit longer on a hot day. The change in volume of solids is too subtle, though, to work in an everyday thermometer. Now that you understand more about liquid thermometers, it's time to get started actually making one!

MATERIALS

- Clear plastic drinking straw
- Ruler
- Fine-tipped permanent marker
- Narrow-necked, small, plastic bottle with lid (Small bottles such as those for medicine, food-coloring or vanilla extract work well.)
- Water
- Rubbing alcohol (Work in a well-ventilated room, and do not leave the bottle of rubbing alcohol uncovered.)
- A few drops or red, blue, or green food coloring (Liquid food coloring works best.)
- Paper towels
- Modeling clay
- Medicine dropper or syringe
- Small bowl
- Hot water
- Ice cubes
- Store-bought thermometer (optional)

PREPARATION

- Empty and wash the small, plastic bottle.
- The clear drinking straw will become the narrow tube of your thermometer. Use a permanent marker to make small marks on the straw, from the top down, at 0.2-inch (0.5 cm) intervals. These marks will serve as level marks on your thermometer.
- Modeling clay will seal the bottle's neck and hold the straw in place. Mold the clay until it feels soft and elastic; then form a ball and push it flat. This round flat piece of clay should be bigger than the neck of your bottle. Use your straw to punch a hole in the middle of this round piece of clay, just big enough to allow a straw to go through. Remove any clay clogging the straw.

PROCEDURE

- Add rubbing alcohol to the bottle, filling it about halfway up. (Reseal the bottle of rubbing alcohol immediately to minimize exposure to fumes and evaporation.)
- Add a couple drops of food coloring to the alcohol, close the bottle and shake it so the liquids mix well. *Why do you think it is important to color the liquid?*
- Fill a medicine dropper or syringe with the colored rubbing alcohol. Carefully set the medicine dropper or syringe aside, preferably on a paper towel in case any liquid leaks out. In case the level of the liquid in your bottle dropped below one-fourth of the way up, add a little more rubbing alcohol to the bottle so the level is one-fourth to halfway up the bottle.
- Poke the straw through the hole in your modeling clay and place the clay on the bottle's neck so the straw hangs in the bottle. Adjust the straw so the end is immersed in the liquid but does not touch the bottom of the bottle. The majority of the straw will be sticking out from the bottle. *Why do you think the straw should not touch the bottom of your bottle?*

- Use the modeling clay to seal the bottle opening and to hold the straw in place. Important: Make sure the clay forms a tight seal around the straw and over the bottle mouth of the bottle, but don't close off the straw's opening. Air should not be able to enter the bottle. *Can you think of a reason why this is so important?*

- Drip the contents of your medicine dropper or syringe—drop by drop—into the straw. *What happens? Does this fluid rise up the straw and stay there?*

- Because no air can flow out of the bottle, the air pressure inside keeps the level of fluid constant and a column of liquid can build up in the straw. Fluid running down the straw into the reservoir indicates a failure of the seal at the top of the bottle. If this happens, make sure the clay forms an airtight seal at the bottle's neck and around the straw and try again.

- The fluid level in the column should reach about midway up the visible part of the straw. If needed, use the dropper or syringe to add more alcohol (uncolored is fine this time) to the straw.

- Observe the fluid level in the straw. This level indicates room temperature. Use the permanent marker to make a small symbol indicating this fluid level on your straw.

- To test the homemade thermometer, hold your hands around the base of the bottle and observe what happens. Be patient, it might take a little time before the fluid in the reservoir adjusts to its new situation. *Does the fluid level in the straw change? Does it change a lot or just a little? Why do you think this happens?*

- *How much will the fluid level change as you heat up the reservoir considerably?* Add some medium-hot tap water in a small bowl and place your thermometer reservoir in the water. Give the thermometer a little time to work. *How does the fluid level in the straw change? Did it change more than when you heated the bottle with your hands?*

- Be sure to pay close attention and take the thermometer out of the water bath if the fluid rises to a level close to the top edge of your straw. *What do you think could cause the fluid to flow over the edge of the straw? What could you change in your thermometer to prevent this from happening?*
- *What do you expect will happen when the reservoir cools down to a temperature below room temperature?*
- Replace the medium-hot water in the bowl with cold water. Add ice cubes to the water so the water cools. Place your thermometer reservoir in the ice-cold water and let the thermometer adjust to its new situation. *What happens? Is it similar or different from what you expected? Could this measurement help you add a scale to your thermometer?*
- Take your thermometer on a tour of your home or yard. Remember to give it time to respond to each new environment, keeping in mind there might not be noticeable differences if the temperature is relatively similar in different places. **Tip:** Look for areas of direct sunlight for warmer readings. *Is one spot the warmest? What about the coolest?*

EXTRA

Make a scale on your thermometer (also called "calibrating" the thermometer) with the help of a store-bought thermometer. *Could you use the fact that in this temperature range, alcohol expands by equal amounts for each gained degree in temperature to make calibration easier?*

EXTRA

You created this thermometer with a particular straw width. Try to build an identical thermometer, now with a narrower or wider straw. *How do you expect the measurements to be different? Will this new thermometer be more or less accurate compared with your first thermometer?*

SCIENCE FAIR IDEA

You used rubbing alcohol for this thermometer. Try to make an identical thermometer, now using water as a fluid. Unlike water, which freezes at 32°F (0°C) and boils at 212°F (100°C), rubbing alcohol freezes at -128°F (-89°C) and boils at 180°F (82.5°C). *Which fluid would you choose to make a thermometer to measure the temperature in your freezer? What about in the kitchen to cook something on the stovetop?*

OBSERVATIONS AND RESULTS

Did you see the fluid rise as you heated the reservoir and fall as you cooled it?

As the temperature of the liquid in the reservoir increases, it expands, and the only place the liquid can go is up the narrow tube. If the temperature of the liquid decreases, it contracts, thereby allowing more liquid to collect in the reservoir. The level in the narrow tube then falls back to a lower level. Once a thermometer is built, each level in the narrow tube corresponds to exactly one volume of liquid, and thus, to exactly one temperature.

A narrower tube (straw) results in a more accurate thermometer because the same expansion or contraction of fluid will cause a rise or fall over a longer distance. As a drawback, the narrow straw might decrease the maximum temperature that the thermometer can reach.

Various liquids can fill the reservoir of a thermometer. How much the liquid expands for each gained degree in temperature as well as its freezing and boiling point are all factors to consider when choosing a fluid for a liquid thermometer.

CLEANUP

To dispose of the rubbing alcohol in the thermometer, pour the alcohol down the sink with plenty of running water. Flush well so no alcohol lingers in the pipe.

Stretch It!

How Does Temperature Affect a Rubber Band?

HERE'S AN ASTOUNDING ACTIVITY TO STRETCH YOUR IDEAS ABOUT PHYSICS.

Have you ever noticed that some objects tend to expand when they get hot and contract when they cool down? For example, you might run hot water over the lid of a jar that's stuck—this causes the lid to expand, making it easier to twist off. Does this effect work the same way for all materials? Try this fun activity to find out!

PROJECT TIME

20 minutes

KEY CONCEPTS

Physics
Heat
Molecules
Forces

BACKGROUND

Materials are made up of atoms and molecules. (Molecules are groups of multiple atoms that are bonded together.) Even though you can't see it with the naked eye, these molecules are constantly vibrating around and bumping into one another. Normally when a material gets hotter, the molecules vibrate more intensely. Because they bump into one another harder this causes the material to expand a bit. This phenomenon is called thermal expansion, which is responsible for many of the cracks you see in roadways and sidewalks. As the temperature changes from very hot to very cold throughout the summer and winter, the materials expand and contract over and over again, eventually leading to cracks.

Rubber bands are different than many other materials because their molecules are actually very long chains, kind of like a tangled pile of spaghetti. These long chains are called polymers, which can behave in surprising ways when you heat or cool them. In this activity, you will change the temperature of a rubber band using hot water and ice water. Do you think a rubber band will expand or contract when it heats up?

MATERIALS

- 2-liter soda bottle
- Permanent marker
- Scissors
- Pencil
- Small, thin rubber band
- About 20 to 25 quarters
- Tape (any kind will work)
- Paper clip
- Hot tap water
- Ice cubes
- Large shallow bowl or pan
- Work area where it's easy to clean spills
- Dish towel or paper towels

PREPARATION

- Make sure you have a work area set up where you can easily clean spilled water, such as a kitchen counter.
- Use adult assistance for cutting the bottle and when using very hot tap water.

PROCEDURE

- Have an adult use scissors to cut the top off a 2-liter soda bottle, turning it into a cylinder shape.
- Use tape to cover any sharp edges around the top rim of the bottle.
- Use adult assistance to poke small holes (just large enough for a pencil to pass through) in opposite sides of the bottle using scissors, about 1 inch (2.5 cm) below the top edge.
- Poke the pencil through one of the holes from the outside of the bottle.
- Loop your small rubber band around the pencil.
- Continue pushing the pencil through the hole on the other side of the bottle.
- Tape together a stack of about 20 quarters, with a paper clip as a "hook" that you can use to hang them from the rubber band. *How many quarters do you think your rubber band will be able to comfortably support?* (The exact number of quarters you need will depend on the size and strength of your rubber band. You want the rubber band to visibly stretch out when you hang the quarters from it but not so much that it hits the bottom of the bottle. A strong rubber band may require more quarters, a weaker rubber band may require fewer.) You can try spreading the quarters out to make them easier to hang—for example, two stacks of 10 taped next to one another or four stacks of five taped next to one another—instead of one big stack of 20 quarters.
- Place your 2-liter bottle into a shallow bowl or pan to catch any water that spills.
- Fill the bottle up to the pencil with hot tap water so the rubber band is completely submerged. Wait a minute or two for the rubber band to come to the same temperature as the water. *What do you think might happen to the length of the rubber band?*

- Look at the bottle from the side so that your eyes are level with the bottom of the rubber band. Use a permanent marker to draw a line on the side of the bottle that is even with the bottom of the rubber band, and label it "hot."

- Now, add ice cubes to the bottle and stir gently. This will cause the water level to rise, so some water will spill out of the holes you poked for the pencil.

- If the water is still very hot, the ice cubes may melt quickly. Continue adding ice cubes and stirring gently until the bottle and water feels very cold. *What do you think will happen to the length of the rubber band at this new temperature?*

- Look at the side of the bottle again, so your eyes are level with the bottom of the rubber band. Draw a new line that is even with the bottom of the rubber band and label it "cold."

- *Is your "cold" line above or below your "hot" line? Is this what you expected to happen? Why or why not?*

EXTRA

You can quantify the activity by using a thermometer to measure the different water temperatures and a ruler to measure the lengths of the rubber band. If you take measurements at multiple points (for example, hot water, room-temperature water, ice water), then you can make a graph of your results.

EXTRA

Try other methods of heating and cooling instead of using hot and cold water. For example, pour the water out of the bottle, and heat the rubber band with a hair dryer. Then to cool it, try leaving the bottle in your refrigerator for about half an hour. *Do you get the same results when you measure the rubber band?*

EXTRA

Try the activity again with different rubber bands. *Do some shrink or lengthen more or less than others? Why do you think that is?*

OBSERVATIONS AND RESULTS

Your results might have surprised you! The rubber band actually expands when it gets colder! This seems counterintuitive because most materials expand when they are heated and contract when they get cold. This occurs because of the unusual polymer structure of rubber. When the long chains get hotter and vibrate, they actually shorten, causing the material to contract. When the chains cool down, they relax and stretch out, causing the material to expand.

To help visualize this, take any rope, string or cable you can find around the house that is a few feet long (short enough that you can hold one end of it in the air without the other end touching the floor). Hold the rope out so it hangs down straight (it is not vibrating) and observe how close the bottom end is to the floor. Now, shake your hand back and forth rapidly (to make the rope vibrate). Does the end of the rope get closer to the floor or farther away? It should get farther away as the rope "bunches up" and vibrates. The same thing happens when the polymer chains in rubber heat up and vibrate—they actually get shorter.

CLEANUP

Pour the water down the drain (or use it—with permission—to water plants) and clean up any spilled water. Remove the pencil and tape from your 2-liter bottle and recycle it—or save it for a future science activity!

Why Does the Floor Feel Cold When the Towel Feels Warm?

CHECK OUT AN ACTIVITY EXPLORING EVERYDAY THERMODYNAMICS.

Thanks to our sense of touch, we can all tell the difference between hot and cold. But does the same temperature always feel the same? Have you ever noticed that the bathroom tile feels very cold to your bare feet but your towel, which has been hanging in the same room, feels warm? Why is there such a difference? Try this fun activity and find out!

PROJECT TIME

20 minutes

KEY CONCEPTS

Temperature
Heat transfer
Thermal sensitivity
Thermodynamics

BACKGROUND

First, we need to consider energy. Heat is a form of energy. When an object gains heat, the object's molecules gain kinetic energy, and they move faster. Temperature is the objective measure of the average speed of the molecules.

Next, we need to consider our sense of touch. Your brain is always sending messages to your hands: "Hey, hands! Grab the doorknob and open the door!" And they return messages, too. Sensory cells in your hands send signals to your brain so it can figure out what your hands are touching. But, as you will find out in this activity, our sensory system doesn't process temperature quite as objectively as a thermometer might.

MATERIALS

- Bathroom with tiles
- Metal towel rod
- Towel
- Empty water glass
- Paper and pencil
- Liquid crystal aquarium fish tank thermometer strip (available at pet stores)

PREPARATION

- Make sure the towel and the glass are in the bathroom you will be using for the activity for at least 10 minutes before you begin the activity.
- Make a list of the items you will be testing: tile, towel rod, towel, glass—and leave space for two notations: how it felt to the touch and what temperature it actually was.

PROCEDURE

- Place your palm on the towel. *Does it feel warm or cold to the touch?*
- Write down how it felt to the touch.
- Now place your palm on the surface of the empty glass. *Does it feel warm or cold?*

- Again, note your observation on paper.
- Continue the same process with the towel rod and the tiles. *How do they feel to the touch?* Note your assessments. *Was it easy to feel a difference in temperature among the different objects?* Wait a few minutes and then hold the thermometer strip to the surface of the towel. *What is the temperature reading?*
- Note this on the paper.
- Now take the temperature of the glass, the towel rod, and the tile. *Are the surfaces the same or different temperatures? How does this compare with the sensation you felt when you touched each object?*

EXTRA

Find three tall glasses. Fill one with hot water, another with ice water, and the third with room-temperature water. Hold onto the ice-cold glass with one hand, making sure your palm is in full contact with the glass. Now hold onto the hot glass in the same manner with the other hand. Keep both hands on the glasses for one minute. Now put them down and immediately grasp the room-temperature glass with both hands. *Is the glass hot or cold?*

OBSERVATIONS AND RESULTS

The objects in your bathroom—the towel, glass, towel rod, and tiles—should all be about the same objective temperature. But the temperature you sense from your hand might be vastly different. This has to do with how heat—or energy—moves. When two objects are at different temperatures, heat flows from the hotter one to the cooler one. The metal towel rod probably felt the coolest because metal is a good heat conductor. That means it will rapidly sap heat from your hand, which makes you sense it as cold. The towel, on the other hand, is made of cotton, which is a poor heat conductor. It doesn't suck away your hand's heat, so it seems warmer to the touch. The glass and the tile might have been somewhere in the middle.

The way we perceive temperature is by comparison. When you touch an object that is a different temperature than your hand, you interpret the flow of heat to or from your hand as the sensation of thermal difference. When heat flows from the skin, its temperature decreases and its sensors "fire" to indicate that the skin is in contact with a cooler object.

Because determining temperature using our sense of touch is comparative, to objectively measure temperature you need to use a thermometer.

If you tried the "Extra" activity, you will see that even your two hands can give you conflicting information if the conditions are right. The hand that held the cold glass told you that the room temperature glass was hot. The hand that had held the hot glass said the room temperature glass was cold. But both hands were holding the same glass—weird!

The mixed-up sensations demonstrate how your hands do not sense the temperature of the glass directly. The heat sensors in your skin sense the temperature change caused by heat—or energy—flowing into or out of your hands. When you held the cool glass, your hand got cold. When this cold hand held the room-temperature glass, heat from the glass flowed into your hand to warm it up. Your cold hand gained heat, so it sensed that the glass was warm. When you held the hot glass, your hand gained heat; when your hot hand held the room-temperature glass, heat from your hand flowed into the glass to cool your hand down. Your hot hand lost heat, so it sensed that the glass was cool.

There are nerves in your hand that can tell if an object is warmer or cooler than body temperature. But they quickly adapt, and the object starts to feel "normal," neither warm nor cold. When you moved your hands to the glass filled with room-temperature water, each hand sensed a change from what it had been feeling. Because each hand was holding a glass of a different temperature previously, the change each felt was different—even though they were touching the some thing.

CLEANUP

Use the water from the glasses to water plants, with permission, or pour it down the drain. Wash and dry glasses. Put away all other materials you used.

Carbonation Countdown

The Effect of Temperature on Reaction Time

TEST THE REACTION TIME OF A COMMON MEDICINE IN THIS FUN ACTIVITY.

PROJECT TIME

30 minutes

KEY CONCEPTS

Chemical reactions
Molecules
Carbonation
Temperature

Have you ever wondered why bubbles form when an Alka-Seltzer tablet is dropped into water? If you've ever tried it, you've seen that the tablet fizzes furiously. The moment the tablet starts dissolving, a chemical reaction occurs that releases carbon dioxide gas. This is what comprises the bubbles. Some factors can change how quickly the carbon dioxide gas is produced, which consequently affect how furiously the tablet fizzes. In this activity, you'll explore whether you can make an Alka-Seltzer tablet fizz faster or slower by changing the water's temperature. How does this affect the reaction?

BACKGROUND

Alka-Seltzer is a medication that works as a pain reliever and an antacid. (Antacids help neutralize stomach acidity, which can cause heartburn.) The pain reliever used is aspirin and the antacid used is baking soda, or sodium bicarbonate. The tablets also include other ingredients, such as citric acid (a weak acid that adds flavor—as well as provides important hydrogen ions, which will come into play as you shall soon see).

To take the tablets, they're fully dissolved in water, where they famously undergo a chemical reaction that produces lots of carbon dioxide bubbles—or fizz. Why is this? As the tablets dissolve, the sodium bicarbonate splits apart to form sodium and bicarbonate ions. The bicarbonate ions react with hydrogen ions from the citric acid to form carbon dioxide gas (and water). This is how the bubbles are made.

How is temperature related to this reaction? For the reaction to occur, the bicarbonate ions must come into contact with the hydrogen ions in just the right way. The probability of the bicarbonate and hydrogen ions doing this is affected by temperature: the higher the temperature, the faster the molecules move; the lower the temperature, the slower they move. (The temperature of a solution is a measure of its molecules' average motion and energy.) Can you guess whether fast-moving molecules or slow-moving ones will speed the reaction time?

MATERIALS

- Two identical jars (You can also use drinking glasses, clear plastic cups, bottles, or vases.)
- Spoon
- Enough ice cubes to fill one of the jars halfway
- Cold tap water
- Hot tap water
- Two Alka-Seltzer tablets
- Timer or clock that shows seconds
- Optional: helper

PREPARATION

- Fill one of the jars halfway with ice cubes. Add cold tap water to about 1 inch (2.5 cm) from the rim. Stir the ice cubes in the jar for about a minute so that the temperature evens out. Right before you start the activity use a spoon to remove the cubes.

- Add hot tap water to the second empty jar until it is about an inch from the rim. Be careful when handling the hot water.
- Continue with the procedure immediately after preparing the jars (so that the water in the jars is still very cold or very hot).

PROCEDURE

- Drop an Alka-Seltzer tablet into the jar with hot water. Time how long it takes for the tablet to disappear. You may want to have a helper time the reaction. *How long does it take the tablet to disappear? How vigorous are the bubbles?*
- Drop an Alka-Seltzer tablet into the jar with the ice-cold water (after having removed the ice cubes with a spoon). Again time how long it takes the tablet to disappear. *How long does it take the tablet to disappear in the colder water?*
- *Do you notice other differences in how the reaction happens in the colder versus in the hotter water?*
- *Why do you think you got the results you did?*

SCIENCE FAIR IDEA

Test Alka-Seltzer tablets in a wider range of temperatures, and then draw a graph showing the time it takes a tablet to dissolve in water at each temperature (check with a thermometer). *What temperature change is required to increase the reaction time by a factor of two (make it twice as fast)? What about decreasing the reaction time by a factor of two?*

EXTRA

Compare whole Alka-Seltzer tablets to pieces of Alka-Seltzer tablets. *If there is a greater surface area (that is, a tablet is broken up into more pieces to expose more surface), and does the same amount of tablet result in the reaction happening faster or slower?*

EXTRA

You can turn this activity into a homemade lava lamp! To do this, you will use an empty container, such as a tall jar or clear plastic 1- or 2-liter bottle. Fill it with about 2 inches (5 cm) of water, add five drops of food coloring, and then fill it at least three quarters full with vegetable oil before adding one quarter of an Alka-Seltzer tablet. You could repeat this activity using your homemade lava lamp at colder and warmer temperatures. (Because it contains oil, you should have an adult help you devise a safe way to warm or cool the contents of each container.) *How does the bicarbonate reaction look in the homemade lava lamp?*

OBSERVATIONS AND RESULTS

Did the Alka-Seltzer tablet dissolve much faster in the hot water compared to the cold? Were there a lot more bubbles produced initially in the hot compared with the cold water?

After the Alka-Seltzer tablet was added to the hot water, the tablet should have quickly dissolved, taking some 20 to 30 seconds to do so, depending on the exact temperature. After the tablet was added to the ice-cold water, it should have taken much longer to dissolve, with most of the tablet disappearing after about two to three minutes, but with some bubbles still apparent after six minutes or longer. In the hot water, the tablet should have more vigorously produced bubbles than in the cold water. The higher the temperature, the faster the molecules move—and the more likely it is that the bicarbonate will contact hydrogen in just the right way for the chemical reaction to occur and produce carbon dioxide bubbles.

CLEANUP

Pour your Alka-Seltzer mixtures down the drain. Wash and dry your containers and put away all materials where you found them.

THE SCIENTIFIC METHOD

The scientific method helps scientists—and students—gather facts to prove whether an idea is true. Using this method, scientists come up with ideas and then test those ideas by observing facts and drawing conclusions. You can use the scientific method to develop and test your own ideas!

Question: What do you want to learn? What problem needs to be solved? Be as specific as possible.
Research: Learn more about your topic and refine your question.
Hypothesis: Form an educated guess about what you think will answer your question. This allows you to make a prediction you can test.
Experiment: Create a test to learn if your hypothesis is correct. Limit the number of variables, or elements of the experiment that could change.
Analysis: Record your observations about the progress and results of your experiment. Then analyze your data to understand what it means.
Conclusion: Review all your data. Did the results of the experiment match the prediction? If so, your hypothesis was correct. If not, your hypothesis may need to be changed.

GLOSSARY

control: The person, group, object, or matter that is not treated with something that is being tested in an experiment in order to allow comparison with a treated person, object, group, or matter.

dissolve: To mix completely into a liquid.

gradient: The rate that something, such as temperature, changes over distance or time.

ion: An atom or group of atoms that has lost or gained at least one electron and thus has a positive or negative charge.

kinetic energy: The energy of movement.

molecule: The smallest amount of a certain kind of matter that has all the features of that kind of matter.

nutrient: Something a living thing needs to grow and stay alive.

obstruct: To block.

particle: A very small piece of something.

phenomenon: A fact or an event that is observed.

prototype: The first model on which later models are based.

radiation: Giving off energy, such as heat. Also, the waves of energy given off.

reaction: A chemical change that occurs when two or more substances combine to form a new substance.

salinity: Having to do with containing salt.

thermal: Having to do with heat.

vibrate: To move rapidly back and forth.

volume: The amount of space an object takes up.

ADDITIONAL RESOURCES

Books

Anduri, Stefanie. *The Environment.* Hollywood, FL: Mason Crest, 2022.

Heinecke, Liz Lee. *Physics for Kids: Science Experiments and Activities Inspired by Awesome Physicists, Past and Present.* Beverly, MA: Quarto Publishing Group USA Inc., 2022.

Noonan, Jim. *Professor Figgy's Weather and Climate Science Lab for Kids: 52 Family-Friendly Activities, Exploring Meteorology, Earth Systems And Climate Change.* Beverly, MA: Quarry, 2022.

Websites

Discovery Education
sciencefaircentral.com

Exploratorium
www.exploratorium.edu/search/science%20fair%20projects

Science Buddies
www.sciencebuddies.org/science-fair-projects/project-ideas/list

Science Fun for Everyone!
www.sciencefun.org/?s=science+fair

Videos

Density and Buoyancy: Mixing Hot and Cold Water
https://ny.pbslearningmedia.org/resource/phy03.sci.phys.descwrld.zhot/density-and-buoyancy-mixing-hot-and-cold-water/

Insulated Mugs | UNC-TV Science
https://ny.pbslearningmedia.org/resource/insulated-mugs-video/unc-tv-science/

The Science of Keeping Cool
https://ny.pbslearningmedia.org/resource/nvmms.sci.phys.cool/the-science-of-keeping-cool/

Publisher's note to educators and parents: Our editors have carefully reviewed these websites to ensure that they are suitable for students. Many websites change frequently, however, and we cannot guarantee that a site's future contents will continue to meet our high standards of quality and educational value. Be advised that students should be closely supervised whenever they access the internet.

INDEX